Organic Food Certification and Marketing Strategies

Gowri Vijayan

All Rights Reserved. No parts of this publication may be reproduced, stored in a retrieval system, or transmitted, in any form or by any means, electronic, mechanical, photocopying, recording, or otherwise, without the prior permission of agrihortico

© 2014 AGRIHORTICO

Preface

This small book is a sincere attempt to provide some basic insights into organic crop production practices and organic certification procedures, especially for those who have no idea about the topic of organic farming. Techniques mentioned in this book are strictly in an Indian context. Production techniques and all other information had been written keeping in mind the present scenario of Indian organic market and crop production practices. However, this book may be used as a reference material for all situations pertaining to organic production technology.

Table of Contents

ORGANIC CERTIFICATION 7
CONCEPT OF CERTIFICATION 7
ORGANIC CERTIFICATION IN INDIA 9
CERTIFICATION PROCESS 10
Situation 1 10
Situation 2 10

Situation 1 10
Step 1 11
Step 2 11
Step 3 11
Step 4 11

Situation 2 11

MISTAKES MADE DURING CERTIFICATION 12
Failure to ask questions regarding certification requirements 12
Missing paperwork deadlines 12
Deliberate use of prohibited substances 13
Forgetting to document the non-GO status 13
Changing of prior approved organic system plan (OSP): 13
Failure to control contamination of field and produce 13
Bad record keeping 14

GOVERNMENT SUPPORT FOR ORGANIC FARMING 14
Support formation or organic farmers` group' 14
Registration of farmers` group with district authorities 14
Documentation of individual farms/farmer's records 15
Service providers 15
Certification and Inspection agencies 15

DOCUMENTS REQUIRED FOR EXPORT ENTRY 15
Schedule B codes 15

Documentation on organic certification 16
Phytosanitary certification 16
Grading and quality standards 16
Contamination and chemical clearance acknowledgement 16

MARKETING STRATEGIES 17

INTRODUCTION 17

FIVE FORCES ANALYSIS OF ORGANIC FOOD INDUSTRY 17

SWOT ANALYSIS OF INDIAN AGRI ORGANIC SECTOR 18
Strengths 18
Weaknesses 18
Opportunities 19
Threats 20

PRODUCT POSITIONING 20

CONSUMER NEEDS 21

MARKETING PLAN 24
Product 24
Price 25
Place 25
Promotion 29

BARRIERS IN INDIAN ORGANIC INDUSTRY 35
BARRIERS IN DOMESTIC MARKET 35
BARRIERS IN EXPORT MARKET 36

ORGANIC CERTIFICATION

Concept of Certification

Organic certification is a certification process for producers of organic food and other products. Certification of any product acknowledges that its production has been done according to organic production standards. The production standards vary from country to country, based on their certifying bodies, but the general concept remain the same:

- No use to human sewage sludge fertilizer in cultivation process

- Avoid synthetic chemical application (See National List, for permitted substances)

- Use of organically converted and certified farms

- Avoid use of genetically modified (G) products in production

- Maintenance of proper records on crop practices, sales and purchase

- Prevent contamination of field and product from non-certified substances

- Use of organic seeds and planting stock, based on availability in market

- Periodic field inspections by certification agents

Also, in order to claim any product as 'organic', certification is required, especially for exporting. Worldwide, inspection and certification of organic foods is done, based on two overlapping sets of guidelines and norms; statutory certification norms and the Voluntary/civil certification norms. The Voluntary/civil certification norms are stricter to statutory certification norms. The Statutory norms are legal guidelines by the government, in relation to certification of organic produce, regulation of import-export, trade agreements between countries etc. Two of the most highly accepted voluntary certification agencies are CODEX and IFOAM

(International Federation of Organic Agriculture Movements). At the international level, the FAO/WHO Codex Alimentarius Commission (the inter-governmental body that sets the food standards) has produced international guidelines for production, processing, labeling and marketing of organically produced foods. These guidelines are followed to, by the members of the Codex Alimentarius Commission. IFOAM also set up an International Basics Standards for Organic Production and Processing, similar to Codex Alimentarius guidelines. Both the guidelines are minimal standards for organic agriculture, intended to guide governments and private certification bodies in standard settings. Governments have used these guidelines for setting up their own organic agriculture programmes. Example is India, USA, and Argentina etc. Some of the other voluntary certification agencies are Naturland, Demeter, soil Association etc

Labeling is an important visual tool of certification. The labels vary the name of the certifying body and the standards with which it complies. The label therein, informs the consumers on the type of standards followed during production and processing of the product as well as the type of recognition granted by the certifying agency. Different organic labels give different meanings. For example, USDA stresses that only foods with 95 to 100% organic can use the USDA organic label.

Labeling protocols followed by USDA for organic products are:

-**100% Organic:** This means the product is or made with 100% organic ingredients, can display USDA seal

-**Organic:** Food contains at least 95% organic ingredients, can display USDA seal

-**Made with organic ingredients**: Product contains at least 70% organic ingredients, cannot use USDA seal, but may list the organic ingredients on packaging

-**Contains organic ingredients***:* Food that contain less than 70% organic contents, not display USDA seal, but may list the organic contents on package

Indian organic labeling specifications also follow similar system for mixed organic products.

ORGANIC CERTIFICATION IN INDIA

Recognizing the importance of organic farming to environment and the economic possibilities from it, Indian government sanctioned the creation of an organic farming intensive policy. In order to standardize the production process so as to encourage production and export of organic products, the National Program on Organic Production (NPOP) was established. The certified products would now be sold under the logo *'India Organic'*. The Ministry of Commerce and Industry spearheaded this movement, with the support of the Commodity Boards. The National Accreditation Policy and Programme (NAPP) was formulated to support the NPOP. According to NAPP statutory certification norm, it is mandatory for all certifying bodies operating in the country, to be accredited by an Accreditation Agency. As of now, there are 6 Accreditation Agencies and 24 Inspection Certification bodies (recognized by APEDA) under NPOP. The 6 agencies are Agricultural & Processed Food Products Export Development Authority (APEDA), Coffee Board, Spices Board, Coconut Development Board, Tea Board, Directorate of Cashew and Cocoa Development. APEDA is involved in global promotion of Indian Organic logo. Export-Import Bank (EXIM) in association with APEDA is engaged in organic agricultural products promotion through awareness creation by participation in International Conferences. The identification of exclusive Agri Export Zones (AEZ) for organic produce in the country, like organic pineapple in Tripura, is being undertaken.

The NPOP standards for production and accreditation have been recognized by the European Commission, Switzerland, and USDA as equivalent to their country standards. This means any product certified according to NPOP can have ready access to European, U.S, Switzerland markets without the need for a separate certification from that country. The trade agreements between U.S, European Union and U.S. and Canada, on organic produce imports and certification, further open this market for Indian organic produce. Approval by National Organic Program (NOP) of USDA opens a lot of markets like Canada and Australia for

Indian organic products. With North America and Europe being two of the top importers of organic produce in the World (2012), this is a highly favorable situation for Indian Organic products to meet their demands. The latest trade agreement between U.S. and Japan on September 26th, 2013, might open the market doors for organic food from countries with certifying agency approval from NOP.

A voluntary certification agency under IFOAM, working in India is the IOAM (Indian Organic Agriculture Movement). IOAM adopted the IFOAM international standards and customized in to Indian farming situation. The farmers growing organic under IAOM standards are eligible to get certificate and the organic label, for use in both domestic and international markets.

CERTIFICATION PROCESS

Certification for a field comes under two situations:

Situation 1
You are going to convert a conventional field into an organic one

Situation 2
You bought an organic certified field from its previous owner

Situation 1
Converting a field requires commitment and proper long term planning. A lot of hurdles have to be crossed, especially if your field is surrounded by conventional farming systems. This requires planning and recording of activities for conversion. It should be noted that practices for soil enrichment, conservation of rain water, adoption of integrated practices should be followed strictly from time of conversion. The certification process considers a 3 year time-period, on an average for complete standardization of the field as per organic standards. The process starts after identification of your certifying agent. The conversion period is calculated on the basis of your date of application to the certifying body or from last date of application of unapproved farm inputs. Remember that clear demarcation between organic and conventional plots is necessary, in case the entire farm is not getting converted together. The certification of processing units is

done once documented evidence on separated processing of organic and conventional produce is submitted.

The certification process can be roughly summed up in 4 steps:

Step 1
Your certification agent will do an initial inspection of the field and submit his report to Certification Review Committee (CRC). Once approval is obtained for the 'go ahead' from the CRC, the farm is placed under supervision for the next 12 months. During this time, any produce from the land, cannot be sold as 'certified organic' or as 'in conversion to organic'.

Step 2
The secondary inspection, done after 12 months will decide the continuity of the project. If satisfied, the farm will be upgraded to 'in conversion'. The farm is to continue this status for the next 2 years, before getting the final certification inspection and approval. The product label used during this period is *'In Conversion to Organic Agriculture'*.

Step 3
This 'conversion period' can be reduced in situation wherein the farmer can demonstrate practice of sustainable techniques, similar to organic standards, immediately preceding conversion, and meet all testing and inspection criteria. The product label used is *'Certified Organic'*.

Step 4
In case the entire farm has not undergone conversion initially, after certification of one part, any remainder of the farm is to be converted to organic within 10 years.

Situation 2
Even if your farm is 'organically certified', the change in ownership status requires you to restart the certification process. You have to find a new certification agent and provide the necessary documents to complete the process 'under your name'. The time taken for this and labeling specifications will be in the hands of your certifying agent.

Certification of farmer groups follow a slightly different procedure, which will be explained by the certifying agency. But, the farmer group should be officially established with the concerned district authority, namely dept. of agriculture, in order to forward the process as a group. Formation of such groups will also help for future risk sharing, reduction in expenditure, better marketing opportunity, buyer-sale price bargaining and other farming processes. An internal control system has to be established, while will be regularly monitored. The success of farmer groups in farming around the world has led to its upgrade to public-private partnerships. The Uttarakhand Organic Commodity Board (UOCB) is one such example of a govt. organization encouraging organic farming. They support the conversion and certification process and provide marketing opportunities to the organic farmers of its state. They also provide training on organic farming and export regulations for them. The complex certification process could be better managed when the farmers are joined in groups, wherein they can share their experience and prevent mistakes from happening.

MISTAKES MADE DURING CERTIFICATION

Certification is an important milestone for a farmer aiming to get certified 'organic'. Yet, many times, mistakes are made due to communication gap, complexity of documentation, length of the process etc. Some of the most common mistakes made are given below:

Failure to ask questions regarding certification requirements
Often, farmers/farmer groups tend to miss out asking important questions on certification process. Documentation is an important part of certification, and many documents need to be collected at the starting of production itself. If unaware, the absence of these documents can cause your certification to be rejected.

Missing paperwork deadlines
-**Failure to pay fees:** Often, certification process involves multiple fee structure, like general organic certification fees, inspection fees etc. The cost may vary for small farmers to large farmers or traders. This also varies from agency to agency. It is essential to

know this, before it is too late. APEDA website may be visited to know the fees structure and other details.

Deliberate use of prohibited substances
-**Mistaken use of prohibited substances:** During the planning stage of production, it is always good to go through the 'National List' of USDA's National Organic Programme (NOP), to see the list of permitted and prohibited substances in certified organic production. Any intentional or un-intentional usage will cost you the certification of your produce.

Forgetting to document the non-GO status
Certification as 'organic' is more of certifying the production and handling process, rather than the product itself. In order to prove the organic nature of your produce, it is necessary to supply documentation of the non-GO (Genetically modified organisms) status of all the inputs and materials used for production. This is very essential, since one of 'the three' taboos to organic farming is use of G products.

Changing of prior approved organic system plan (OSP):
Every farm has an OSP submitted to the certifying agent at the beginning of the conversion process. This record helps the agency to checklist the progress, during inspections. The farmer is allowed to make changes in his OSP, but only after duly informing his certification agent of the changes made. This will help in avoiding any confusion made during the next inspection.

Failure to control contamination of field and produce
One of the most common hurdles to organic farming is fear of contamination. If your neighboring farms are practicing inorganic or using genetically modified (G) products, it is necessary to buffer your field from them. Also, contamination during postharvest handling is a common incidence. This is attributed to use of same harvesting machinery for both organic and conventional products. In order to certify your field as organic, it is essential to follow proper buffering techniques, placement of no-spray signs near your operation area, implement a GO drift management plan etc.

Bad record keeping

A detailed recording of the production plan and crop schedules is to be strictly maintained. Documents like field maps documenting the acres, borders, field numbers, adjoining land users, receipts from sales and purchases, service records of plating, spraying, harvesting, processing and delivery, harvest and storage records, lot numbering system are essential during certification procedure.

GOVERNMENT SUPPORT FOR ORGANIC FARMING

Along with establishment of NPOP certification and APEDA branding procedures, the Ministry of Agriculture is also promoting organic farming under the National Project on Organic Farming, National Horticulture Mission, Technology Mission for NE and Rashtriya Krishi Vikas Yojana. The National Project on Organic Farming supports training on resource management, technology development, information dissemination, market development and awareness. The Rashtriya Krishi Vikas Yojana assists states in area expansion under organic crops, training of farmers and production of organic inputs. The National Horticulture Mission and Technology Mission for North-East provides financial assistance to organic cultivation at the rate of 50% of cost, to a maximum of Rs. 10,000 per hectare (maximum 4 hectares per beneficiary). Financial assistance for construction of vermin-compost units and farming certification (farmer groups) is also provided.

In order to encourage genuine organic farmers and to increase their business opportunities, the Govt. of India and State governments have initiated several steps. A few among them are:

Support formation or organic farmers' group'

Such groups can help each other in matters of inspection of fields, conversion process, certification, monitoring and supervision over the practices. Also, marketing of the organic produce could be executed effectively in numbers.

Registration of farmers' group with district authorities

Registration could be done under Dept. of Agriculture of Dept. of Horticulture, as per produce details. This would cover information on the individual members, plot numbers, area and crops planned etc.

Documentation of individual farms/farmer's records
This is very important for later certification procedures. In case the farmer is unable or unaware of proper recording procedures, service providers registered with the State government may provide the same under nominal fees.

Service providers
State governments register several agencies to support farmers in production and marketing by disseminating knowledge and providing basic facilities. Some of those agencies are Krishi Vigyan Kendras (KVKs), State Agricultural Universities, Agri Clinics, NGOs, private entrepreneurs, Central agencies etc. Organic farmers can get assistance from these offices, regarding organic cultivation practices and recording of data.

Certification and Inspection agencies
Govt. Of India has 6 organic certification bodies to manage certification procedures for organic cultivation in the country. APEDA, one of the six certified bodies, has identified 24 certification agencies to process the certification procedures throughout the country. The current licensed agencies could be confirmed from NPOP website http://www.apeda.gov.in/apedawebsite/organic/.

DOCUMENTS REQUIRED FOR EXPORT ENTRY
The following documents are required while exporting organic produce to another country:

Schedule B codes
These are required when submitting either the electronic Automated Export system (AES) or Shipper's Export Declarations (SED) version of export shipment data. This has been established in order to facilitate tracking of imported and exported organic products throughout the world.

Documentation on organic certification

The product organic certification agency authority, will be confirmed, to meet the importing country standards.

Phytosanitary certification

In general, a phytosanitary certificate issued by an official of the exporting country is required to be presented at the import entry point (Ship dock). This will be verified and confirmed by the concerned authority of the country.

Grading and quality standards

Exported fruits and vegetables are required to meet the import requirements related to the size, grade, quality, maturity.

Contamination and chemical clearance acknowledgement

While the first 2 documents are specific for organic products, the last 3 documents are same for any imported product along with import clearance.

MARKETING STRATEGIES

INTRODUCTION

Marketing strategies are based on the needs of the business. Planning of market strategy for organic farming in India requires a clear understanding of the present conditions of the industry. Strategies are based on selection of product, type of market, recognition of consumer needs, industry characteristics, price, marketing channels and promotional strategies. In the Indian context, the absence of a stable domestic market for organic food makes it necessary to concentrate on market confirmation/establishment. It is thereby essential to understand the present situation of the market, its preferences, competition, replacements and entry barriers among other issues. Michael Porter's five forces analysis of the organic food industry and SWOT analysis of Indian agri-organic sector are given below.

FIVE FORCES ANALYSIS OF ORGANIC FOOD INDUSTRY

Bargaining power of suppliers: High concentration of suppliers for organic inputs

Bargaining power of customers: End point consumers are price sensitive

- Customization of produce for buyers
- Low dependency on distributors

Intensity of existing rivalry: High growth rate of organic food industry

Threat of substitutes: Conventional food prices are less

- More number of substitutes in market
- Familiarization of buyers with existing conventional products
- High product differentiation and branding of substitute products

Threat of competitors: Capital requirements (high)

- Need for establishment of strong organic food supply chain
- Entry barriers (e.g. certification)

SWOT ANALYSIS OF INDIAN AGRI ORGANIC SECTOR

Strengths

Wide diversity in agro-climates across the country, supports wide range of agri. and hort. crop production

Seasonality does not limit the availability of any produce due to different climatic conditions in different parts of the country. This will support export of food produce throughout the year, irrespective of season.

Traditionally, farming systems in India use fewer pesticides. This will aid in easy conversion of such fields into organic.

There is increasing interests among farmers, entrepreneurs and govt. in organic farming, with various agriculture universities bringing forward various organic crop production POPs (Package-of-practices).

Indian corporate firms are increasing investments in agribusiness and specialty farming like organic farming.

Differentiation can be created easily based on product nature and background

Weaknesses

Reduced shelf-life varieties prevent successful sale of produce overseas

Lack of awareness of organic practices and certification needs and requirements pulls back growth of organic industry

Price decision is an issue with Indian organic food exporters. They face problems in international markets against organic produce from other countries. International Organic farmers supported by subsidies from their respective governments, quote lower prices and this pose a problem to Indian exporters.

Lack of sufficient MIS (Market Information System) prevents planning of successful marketing techniques and sales opportunities

Low quality post-harvest handling and processing techniques and management prevents Indian organic products from getting good price in markets

Insufficient R&D in areas of organic packaging, handling and processing

Lack of awareness of certification and labeling specifications for export of organic produce often leads to return of products

Previous quality restriction cases on Indian food products close many international markets to remaining organic agro-products.

Opportunities

Support of government in organic farming and export

Opportunities being offered by WTO in organic farming and global trade

Premium price for organic food in international markets

Increasing demands for organic food in international markets, particularly middle-eastern countries

Initiatives taken on branding Indian organic produce by APEDA

Establishment of big retail outlets and chains in the country offer future prospects in demand hike for organic produce

Interests shown by private sector to involve in organic food supply chain offers wide range of possibilities for improvement in Indian organic market performance

Threats

Competition from the conventional products in the domestic markets

Threat from imported products

Barriers on trade and specific certification specifications by international markets

PRODUCT POSITIONING

Studies on performance of Organic food and food products in the Indian market show decrease in sales from all quarters. The decrease in sales has forced several retailers like ITC to withdraw a portion of their organic food profiles from the markets. Retailers attribute this to less awareness among consumers regarding organic foods. Consumers assume organic food to be of less quality when compared to their substitutes, particularly based on the color of produce. For example, the color of organic turmeric being less yellow or Chili powder less red leads to such conclusions. Also, the high premium price of about 50-70% above conventional productions, further divides the consumer interests to organic foods. The fact that only a part of their diet will be of organic origin seems impractical for consumers to pursue further. Therefore, it can be concluded that there are two major problems facing organic farmers and retailers: High cost of produce and perceived low quality of the produce. The farmer groups have to consider both the decisions before deciding the positioning of produce. Based on the perceptual map of positioning, there are four situations for any product to position itself:

Position 1: High quality and high priced product

Position 2: High quality and Low price product

Position 3: Low quality and low price product

Position 4: Low quality and low priced product

The positioning of the product depends on the strategy selected. Porter gives us three competitive strategies to select while developing a market for product namely cost leadership strategy, differentiation strategy and niche strategy. Indian consumers are concerned with price and this overrides their environment and health commitment consciousness. Also, quality confirmation is needed in this case for encouraging increase in sales. Without compromising on quality, the farmer groups can only focus on position 1 and 2. If proper differentiation of products and services is done, position 1 could be maintained, but if cost management strategy is followed position 2 could be achieved and be more desirable

CONSUMER NEEDS

A basic idea of types of consumers is essential to understand the marketing strategies to be adopted. Consumers have been classified into five categories based on their inclination towards a product. The first stage of 'suspect' refers to a person who comes in contact with the product information for the first time. It is the business of marketers to ensure public cover this level, so that the desirable levels can be achieved. The next stage is the 'prospect'. Marketers see all persons as prospects, who have potential to develop into the next level i.e. 'customers'. The next two levels are of great importance. Customers have the potential to upgrade into 'clients' who like some aspect of the product. Clients change into 'advocates', the highly desirable type of consumers. They not only are loyal, but also self- promote the products by word of mouth. Indian consumers have a general idea of organic products and its availability. The task for the farmers/farmer groups here is to create prospects and maintain customers. Through creation of advocates is the ultimate goal, under present conditions of Indian organic food industry, it is essential to convert more customers to the organic style. Identification of basic needs of consumers is essential to further guide farmers through the positioning and planning of products.

According to Maslow's hierarchy of needs, consumer needs are categorized into five types. Physiological needs are the basic needs for food, water and shelter. This lead to safety needs of concern for health, environment and other physical and emotional requirements. The social needs of friendship, belonging, love follow. The next level is esteem needs of self-respect, status and recognition. The last level is of self-actualization with need for personal achievement, reaching potential and exceling in life. The positioning of organic product with intention of fulfilling physiological needs would be difficult in the Indian context. With the wide number of substitutes available, promotion of high priced organic produce for basic need fulfilment is not going to work. The basic point is to bring up the level of awareness of consumers step by step. Indian consumers are weary of buying organic food, due to some misconceptions and price hindrance. Therefore, the safety factor of organic food consumption has to be brought up first. The farmers have to tap into the sales potential to the Indian middle class. Under present circumstances this could only be achieved by marketing based on safety and social needs of healthy food, free of pesticides, locally grown, a feeling of community. Basically, marketers for Indian organic food have to lure prospects through promotions based on safety needs, convert them into customers and maintain them through social need actualization. Over time, with development of domestic market, the positioning can be changed to upper levels. A pictorial representation of the discussion is given below. In case of high end/upper consumers, esteem needs play a very important role. For them, the marketers have to concentrate on promotions based on esteem needs of societal acceptance and presumed support to environment conservation.

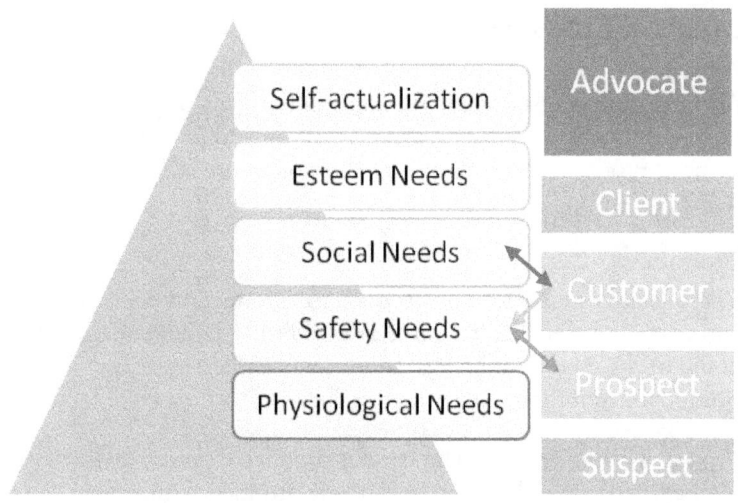

Maslow's Hierarchy of Needs   Types of consumers

MARKETING PLAN

A marketing plan is basically based on the traditional P's of planning. In case of organic farming, all the 4 P's play equally important roles to capture market attention and interest. The 4 P's namely; Product, Price, Place and Promotion are explained in detail below.

Product

The selection of the type of produce for production depends on various parameters like resource availability and market demand. Some of the factors to be looked into while planning for production and marketing are given below:

Product Characteristics- The type of produce required by the markets is to be understood before any production planning. This will help with marketing of its desirable characteristics. The size of the produce, the color and flavor favored, freshness required of the produce has to be understood before going into the production process as such.

Crop variety- Often, corporate buyers prefer a particular variety of the crop to be available. Identification of such variety requirements will aid in getting prospective buyers and support contract farming.

Environmental concern- An environmental commitment angle will help the farmers in further selling their produce, using environmental cause as a stepping stone.

Personal touch- Production of a local special is another way of promoting the community. This could be taken into consideration for getting community support

Seasonality demands- Seasonality demands for different produce in different countries have to be taken into account while produce planning, especially if interested in export.

Packaging- Updating on consumer requirements for packaging, latest technology in packaging will help farmers to better present their produce and interest retailers.

Price

Pricing is a very crucial element of marketing strategy. The high price associated with Indian organic products is a bane to its acceptance in the domestic and international markets. As per recent studies, Indian consumer consumption of organic food is very low. This is attributed to lack of awareness and high price of the product. The large number of middle class and upper middle class members of the Indian society have to be attracted to organic produce, in order to develop the local market. In case of exports, the high price quoted by Indian exporters often draws less interest and business from international buyers. This is also supported by the low quality of produce offered for the high price quoted. In order to expand the Indian export market, it is necessary to upgrade the quality and quotation of realistic and viable prices. The different pricing strategies are:

Static pricing-The price will be same for all groups of customers

Flexible pricing-The fixed and variable costs of production and export are covered by the price

Marginal pricing-When the market is on its 2nd stage of development, the pricing could be quoted considering only the variable costs of production and export

Penetration pricing-In order to attract more customers, the price could be quoted quite low, thereby discouraging competitors, gaining market share and profit through numbers

Market skimming-In case of very little competition, a premium pricing of the product could be settled for to make profit from high-end consumers

Place

Place or more specifically marketing channels define reach of the produce into the market. In majority of cases on organic farming across the world, market initiation has been done either by an entrepreneur or exporter or business or any government agency. The Uttarakhand Organic Commodity Board (UOCB) of India is such an example. Indian organic farmers/farmer groups have three market options namely, domestic consumer markets, domestic business markets and export market. The channels used for each market differs according to the need of the buyer.

<u>Domestic consumer markets</u>- Such markets bring farmers in close contact with consumers. It is essential thereby, to understand the consumer preferences and inhibitions. Considering the large number of potential buyers, the farmers have a real good chance to make huge profits, if channeled well. An understanding of the purchasing pattern of consumers and preferences in shopping is to be taken into account for channel selection. The first thing to ensure here is consumer accessibility of product and product information on all levels. One of the most important problems faced by working individuals is time constraint. Strategies could be placed on time saving, in order to attract such consumers. Traditionally, farmers come in direct contact with consumers through farmer markets. Such farmer markets are limited to once or twice a week, resulting in time and accessibility constraints. The opening of a farm gate store would not only increase chances of increased sales but also remove the time constraint for consumers to shop. As an added initiative, home delivery service could be offered for regular customers, further saving their time. This could be used as a loyalty incentive. One successful marketing model adopted by western countries like United States is community supported agriculture (CSA). CSA is basically an association of consumers and local farms. There is direct interaction between the farmers and consumers without any middle men. The consumers pledge to support one or more local farms, sharing the risks and benefits of production. The CSA members or subscribers pre-book the anticipated harvest by paying for it at the starting of the season. Once harvesting completes, the vegetables and fruits are send to the subscribers on a weekly basis. The share of produce is distributed in box formats to the subscribers. This model not only takes care of the accessibility and time constraints, but also gives a sense of ownership and local commitment to the consumers. This model works on mutual cooperation and understanding between the concerning parties. While the farmers should be aware of local community requirements, the subscribers should also express their needs and financial limits to the farmers. CSA model is widely used in many countries especially for organic produce, and is yet to develop in India. The development of such a model will assure

continuous payment for organic farmers without any intermediaries and other problems associated with them. In order to cater to more customers, it is necessary to expand the product reach further. Internet is a source of entertainment, education and more important shopping for consumers. This consumer tendency of internet surfing could be used to advantage. A website on the farm containing information on products available, services provided, prices and promotion could be made available. The promotion of this website could lead to increased viewing and consumer update on produce availability. Considering the increasing penchant for online shopping, providing options for online shopping of organic produce is a logical strategy. Countries like Brazil and Hungary already follow this system for their organic produce.

Domestic business markets- The domestic business trade caters to the needs of retailers, wholesalers, processors, restaurants and health stores. Institutional sales of food produce are practiced in Brazil, but might pose a problem in India due to price constraints. The chances for high profits from this segment are usually less. Contract farming and direct selling to buyers are two of the common methods of distribution. Irrespective of the method, it should be remembered that in this type of channeling, the availability of specific variety produce of specific quality and grade is very essential. Timely supply of specified quantity of produce is a rule for these buyers. Entering into contact farming reduces the produce selection options for farmers and produce specifications have to be followed to the point. Risk coverage is often involved in the contract and local biodiversity could be affected by planting of exotic varieties. Direct selling to the buyers involve price quoting and auctioning. Price is a very important point of consideration here. All market demands on quality, variety and grade have to be understood before selection of channel.

Export market- The formation of farmer groups and co-operatives is recommended especially while planning for export. This will help in certification process, sharing of risks, selection and bargaining with buyer. Farmers/farmer groups can either opt for selling to business or directly selling in the international market. Farmers producing for export could supply to an exporting company, who will look further into the processing and packaging of the produce. Farmers looking for exporting of their produce could join in a private-public partnership to utilize available facilities and get market exposure. The UOCB is one such government organization which supports such partnership with organic farmers in Uttarakhand. Along with market initiation, UOCB not only identifies the buyers but also help in bargaining by farmer/ farmer groups. The UOCB also trains the farmers in organic farming and provides a common facility center for organic produce (CFC) for post-harvest processes. Organic farmer cooperatives across the world are also known to form Limited company (Ltd.) in order to self-export their produce, for example the African Organic Producers Co-operative. Indian farmers/farmer groups could use export agents to support the export of their produce. Hungary Organic Food Producers also follow this system for facilitating exporting. The farmers could also tie up with an entrepreneur who exports organic food (e.g. African Organic Fruit Industry).

Summary of the above discussion on marketing channels is given below:

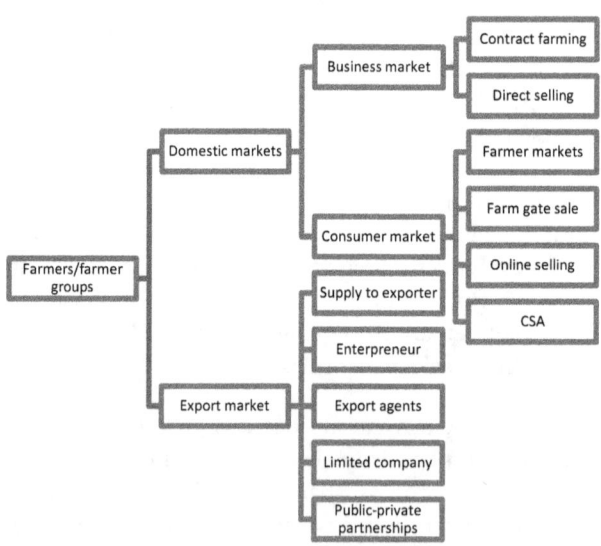

Promotion

Promotion of Indian organic food could be split into promotional activities for local and export markets. Promotional activities on both instances need to focus on branding and visibility of the products.

Promotional activities for international market-Lack of branding of Indian organic produce in the international markets reduce chances for good price for Indian produce in the market. Understanding this, some of the government organizations have come forward with branding initiatives. APEDA has taken up the responsibility of creating brand awareness of 'India Organic', the NPOP certified organic logo for export. Also, UCOB has taken up branding of organic produce from Uttarakhand. Majority of the Indian farmers sell their produce to exporting businesses and therefore are not concerned with international market promotion. However, farmers do have the option of directly selling their produce, as explained earlier. Such farmers will have to focus on promoting their farms and marketing/distribution capability. The formation of farmer co-operatives or Limited company (Ltd.) are preferred arrangements in this situation. It will provide the necessary quantity of produce, of assured quality, ease distribution and marketing and also share risks involved. The following suggestions might help farmer groups/co-operative in brand promotion in international markets:

Creation of a brand-Selection of logo for representing the farm/co-operative goes a long way to product recognition

Participation- The farmers/farmer groups could participate in international trade fairs and exhibitions to understand the current scenario and even open a booth for their products.

Website-A website for a farm/co-operative not only creates awareness but only gives the necessary authenticity to the farm. Any useful video clippings and articles on organic farming and the farm could be placed to attract buyers and consumers. Social networking sites are great for capturing the local consumer network, rather than international business firms. A LinkedIn profile could be created for the firm, the link to which could be promoted through the official website of the farm.

Promo-aids-Pamphlets and catalogues on farm's products, prices and other details could be distributed during such trade fairs and exhibitions. It is the best ground to promote the farm logo through these information aids. Information of farm website can be also provided so that interested can look into details of the farm online.

Meetings- Trade fairs and exhibitions acts as places to meet important persons in the industry. The farmers/ representative can set up a meeting with some international importers and retail businesses. They can be invited to your farms, for observing the practices followed. If found satisfactory, further talks on sales could be done.

Visit to markets-Though expensive, a visit to the local markets while attending international exhibitions will give an idea of imports, their pricing and problems. This could be taken into consideration while meeting with the local authorities for discussion during the talks. Any interests shown towards local market development might even result in an invitation for business in that country.

International agencies- The farmers can tie up with IFOAM or IOAM for conducting any activities in their farms. This could be further developed for collaboration with farms from other countries, like an experience sharing activity. Such interaction with international agencies will not only promote brand of the farm, but also place it in the international scenario. It would also expose the farmers to practices and knowledge of farmers from other countries.

Tie-up with govt. organizations- Farmer groups and cooperative could tie up with the APEDA and the 6 other commodity boards for their branding activities. This will reduce the weight of self-branding from the farmers and also give sufficient market exposure.

Tie-up with export company- In case of marketing difficulty, farmers could tie up with the best possible exporting firm, in order to use their contacts for market establishment.

Magazines/Journals-Though newspapers are more wide spread, the required technical audience is limited to journal and technical magazine readers. Studies on farm performance or any technical feature could be conducted in the farm, with professional support and published in prominent international magazines or journals. The stress on farm logo and idealistic organic practices followed, could help create interest among the readers. Such publication information can also be displayed in the farm website, to showcase commitment to research. This will also give more authentication to farm claims on produce quality and services.

Promotional activities for local market- Indian domestic market for organic food faces two main problems namely lack of awareness about products among consumers and high price tension. While the pricing decision depends on the strategy to be adopted for market entry and development, awareness creation solely depends on promotion. Awareness of the product can be improved by product recall and consumer education. Efforts to ensure the visibility of the produce is necessary to affect product recall. Since the local market buyers consists of consumers, retailers, wholesalers, processing industries, restaurants and health stores, divisional promotion is required for the groups. Push strategy is adopted against the business buyers (retailers, wholesalers, restaurants etc.). It should be always remembered that farmer-business ventures require strict follow up of product specifications given by the buyer, along with the time limit. Some of the suggestions for use are:

Discounts-Bulk buy discounts could be offered for attracting buyers

Representation at trade shows-It will give the necessary opportunity to interact with the business representatives to set up business meetings and even exhibit farms produce.

Use of mail-Mail shots could be send to the distribution chain as a source of business proposition.

Farm visit- The representatives of businesses could be taken to the farm to see the production, processing facilities to assure claims on product quality. It would be very beneficial to know about the business details, their product requirements like variety, quantity needs, quality requirements, before talking to its representatives regarding possible business. Businesses might consider a well-informed farmer/farmer group to be a desirable trait.

Creating visibility among consumers takes more time and more innovativeness. The pull strategy is adopted to attract consumers to the farm products. Some of the most common practices followed include distribution of demand-discount coupons, promotional schemes, loyalty schemes and product quality testing sessions for consumers. Though these measures ensure some level of consumer movement, under Indian context, increasing visibility and awareness are also essential. Some suggestions for increasing awareness and visibility are given below.

Email marketing- Regular customers could be updated on the produce availability, price and other details through direct email. This is useful particularly during CSA form of farming, wherein the subscribers can be regularly update on harvest situation.

Web public relation (WPR)-A website of the farm, with information on product profiles, seasons of availability, services rendered by the farm, prices and other relevant information could be displayed. Besides being a source of communication over distances, websites also prove authenticity of the farm's claims. Though many websites of illicit nature are also present online, people expect established businesses to have a website with all essential information. Any news worthy stories on the farm products or services, pictures of the farm and products, customer review and feedbacks could be placed in the website for prospective customer review. Display ads like pop ups, videos, floating units and tactic banners and payment to search engines for placement of farm website among its listings, wherein a potential customers enters a relevant search term are some of the other tactics of WPR for increasing customer recall rate..

Social media networking- The sites provide opportunities for advertising of products (paid). Farmers could use this opportunity to promote their farm products and services, highlight environmental or any CSR features of the business. Various farming games like Farmville and Farm town of Facebook converts curiosity of users into entertainment. With the help of web technical experts, farmers could create and upload an innovative interactive advertisement of their farm, taking social networking user profiles into consideration. The more innovative and refreshing the advertisement the more word of mouth for the farm. Professional sites like LinkedIn could be used to attract students. A LinkedIn profile could be created for the firm, the link to which could be promoted through the official website of the farm. F&Q sessions could be arranged with interested students and professionals in agriculture and agribusiness through LinkedIn profile. This would provide an opportunity to sharing experience, gaining new knowledge and spreading the farm name. The LinkedIn account link should be included in all the promotional materials distributed by the farm. This will expand the levels of communication channels open for suspect and prospect customers.

Community/ grassroot marketing- This is especially useful for CSA model of organic farming. It involves door-to-door distribution of flyers and leaflets on the farm and produce details. This method is used generally for tapping the local community/ catchment.

Internship-Student internship programs could be organized with the local agriculture university and colleges. This is being practiced in western counties and bring good results. This method will not only increase the awareness level of students and spread by word of mouth, it would also bring the farmers in close working with the government officials and researchers. This could be used to advantage of the farmers in business creation. Also, any articles on the program could be sending to newspapers for publishing, further promoting the farm.

Media marketing and public relation (PR) - Media and PR have a close relationship. The farm could do a write-up on any topics on important days like World environment day, Earth day etc. and publish in newspaper/magazines. Response generating advertisements could be placed in newspapers and magazines. The farm could also organize any event which would get media attention and get it published. All these could provide necessary spread for the farm. Press releases on farm activities and farm literature are other PR activities recommended.

Agri-tourism- Agri-tourism is practiced in parts of the country. It is one of the innovative methods of promotion widely adopted by farms of the western lands. Agri-tourism involves invitation of customers to visit farms during a particular agricultural operations for example, harvesting drives are organized wherein people harvest the crops. Agri-tourism often involves farm stays. This method aims at rekindling public interests in crop production practices. It also ensures the following of healthy and sanitary production measures by the farm. Consumer awareness could be increased and misconceptions on organic products removed at the same time.

Local conferences/fairs-Participation in local conferences and fairs like KrishiMela will provide the necessary exposure of participants to organic farming in the community. It will also give a chance for farmers to interact with interested students, researchers and government officials. The high number of local participation in such melas makes it the ideal event for mass scale promotion of the farm and organic products.

Information aids- Information aids like fliers, pamphlets and catalogues on farm and produce profiles could be distributed during fairs and conferences. The more number of farm materials spread, the likely chance of increased recall by people. The intention is to spread the farm name and logo so that recall leads to interest into buying intension and then buying by consumers.

The limit of all promotional ideas and tactics are limited to catching attention. The rest of the work depends on the product and the distribution system planned, among other parameters. In order to succeed in organic farming, the farmers have to understand the barriers in the Indian organic industry, both domestic and export. The next section highlights some of the barriers identified in this industry. Removal of all barriers is essential for the establishment of a successful and viable market for Indian organic products.

BARRIERS IN INDIAN ORGANIC INDUSTRY

Both the Indian domestic and export markets face different barriers to development. These hurdles have to be understood and resolved in order to capture respective markets. The barriers are briefly explained below.

BARRIERS IN DOMESTIC MARKET

Indian domestic market is not as developed as export segment. This is linked to a serious of hurdles in the industry, which are listed below:

Price: Organic produce in the local market, costs 40-60% more than their conventional counterparts. Presently, domestic organic market caters to the needs of a very limited percentage from high income group, leaving a huge gap in consumer capture. Price reduction or value addition, keeping in mind the lower groups of consumers is necessary to attract them into this segment.

Awareness: Ignorance of organic farming and its certification procedures prevent production of food products qualified to be certified 'organic'. The help of NGOs and other groups can be taken up to spread the information and facilities to farmers, spread across the country.

Government involvement: The government agencies concentrate on exports of organic products, thereby ignoring the domestic market. The absence of subsidies for organic farmers (local market) further prevents pursuit of organic farming for the local markets. Lack of market creation, infrastructural support and marketing further disappoints the organic farming sector.

NGOs: NGOs play an important role in spreading awareness regarding organic cultivation and marketing among farmers and consumers. The lack of working linkage between NGOs and certifying agencies, funds and infrastructure prevents successful organic farming training and implementation.

Traders (domestic): the low level of awareness of the product, low price realization and lack of proper marketing network prevents steady market creation for organic produce in India. Considering traders to be an important link between the producers and consumers, it is necessary to focus on their performance enhancement.

Farmer groups: Organic farming is resource intensive and requires the back-up of money and good marketing channels to be profitable. This has led to formation of strong organic producer organizations and farmer groups in the West. In this case, the lack of modern technology to upgrade the production process and less focus on domestic market demands makes its development difficult. It is necessary to encourage the farmers to concentrate on the demands of the local market too. This can be done by the government, through schemes and subsidies focusing on domestic markets, providing more marketing opportunities etc.

Promotion: There is a lack of promotion of organic foods in the local market. Organic foods, which are costly, share the market with cheaper, conventional products. In order to create more consumer base, it is necessary to reach out to all segments of consumers.

BARRIERS IN EXPORT MARKET

The export market has proven tough on Indian organic products. Several concerns and issues prevent maximum development of this sector. They are:

High price expectation: the high price expectation by Indian producers prevents sales in the International market. Considering the quality issues preceding Indian products and the stringent International regulations, it is highly advisable for Indian Organic food exporters to reduce their prices as per situation and requirements.

Quality: Issues of quality and contamination prevents entry of produce into many markets. This has to be taken into consideration by the government and APEDA and suitable technology provided to solve the issue.

Marketing: Better promotion of products need to be don't in the international market, by exporters as well as government institutions.

Logistics: Shipments issues and delays due to shipping documentation prevents easy movement of produce

Certification: Confusion on certification procedures and lack of documents prevents proper certification for exporting of produce.

Export authorities: Complicated paper work delays the process. It is necessary for the authorities to find an easy way of documentation, preferable online.

Information: Difficulty in attaining International market requirement news and certification policies, prevents suitable planning of business. The establishment of any easy access portal for the same, will save a lot of time for the producers and exporters, to plan their business.

www.ingramcontent.com/pod-product-compliance
Lightning Source LLC
Chambersburg PA
CBHW051826170526
45167CB00005B/2169